Particle Level Chemical Engineering CAD Using Arno Vigen Scrunched Cube (AVSC)

Empowering Chemical Engineering and Nanotechnology

By Arno Vigen

© 2018 Eric Arno Vigen

Simple Words to Understand . . . Chemistry, Elements, and Bonds

Why does a Nucleus Stay Together If Protons Repel?

- A Nucleus is Just . . . a Magnetic Ring

Electron Shell Chemistry Is Just . . . Scrunched Cube Geometry

- Why are electron shells in sets of 2, then 8, then 8 and such? Can we improve Pauli-aufbau?

Scrunched Cube Electron Shell and Bonding Periodic Chart of Elements

- Understanding Molecular Bonding in the Scrunched Cube Atomic Model

Scrunched Cube Molecular Bonding

- Understanding Molecular Bonding in the Scrunched Cube Atomic Model

Scrunched Cube Explains in 3D What Makes Molecules Solid, Liquid, or Gas

- And Why is Gas of Every Element the Same Volume (a mole)?

Particle Level Chemical Engineering CAD Using Arno Vigen Scrunched Cube (AVSC)

- Training Course for SolidWorks Engineering Tools

Simple Words to Understand . . . Gravity and Other Forces

Gravity is Just . . . That Electrons are a Little Closer

- Explaining Gravity from the basics of Electromagnetism and Explaining Why Observed Mass Changes

The Five Continuous Fundamental Electromagnetic Forces: Reconnecting Newton into the Chemistry and Particle Physics

- Resolving Strong Force, Weak Force, Bonding Force, and Gravity via the basics of Electromagnetism as One Continuous Function

Fixing Einstein's E=mc-Squared

- Integrating that Mass is Magnetics divided by the Electron Shell Radius Volume Explains

Solving Schrodinger's Equation

- Using Intermittently Expressed Gated Accumulates of Nucleomagnetics Potential

How is Electricity and Magnetism Linked?

- Exploring the Fundamental Linkage of Charge and Magnetism

Simple Words to Understand . . . Personality

Visual Astrology: Fun, Support, Security, and Growth

- Astrology 'signs' archetypes are based upon powerful traits to understand people

Visual Astrology Relationships

- What happens when 'sign' personalities interact

Visual Astrology and Jung

- Astrology 'signs' archetypes actually predict all the Jungian 4 archetypes

Dominant Personality Traits

- Dominant Personality Traits Follow from Four Dimensions, Six Steps and so 24 Subcategories

Simple Words to Understand . . . Communications

Decision Matrix® Writing

- Persuasion is based making arguments and comparisons at the correct strength in an understandable, powerful order.

GATESOUP® Writing

- **G**oal, **A**udience, **T**heme, Enough **E**lements, **S**upport and the rest

Kedarf® Grammar and Composition Explained

- Defining the Parts of Speech, Paragraph Structure and More in Usable Terms

Table of Contents

Electrons, for Most Engineering, Have Settling Positions 7

    3D Structures Build in Two Hemispheres of Settling Positions 1/3/5/7/7/5/3/1 from the Axis to the Equator to Axis 13

    However, 2nd Layer Can Fit In Between One Layer to Create Repeats 1/3/3/5/5/7/7 17

    Shell-2 and Shell-3 Scrunched Cube 19

    Equatorial Exceptions to Shell Filling 21

In AVSC Chemical Engineering Tools, Electrons (and Protons) have Repulsion Zones 23

    Special Feature – Adjusting Repulsion Zones as a Set 25

    In AVSC Tools, Repulsion Zones Size Based upon the Distance (Layers) from the Exterior 30

    Protons Have Repulsion Zones 32

An Atom (Each Nucleus) Has Bond in Open Paths past the Electron Field into the Nucleus as Specific Inclination/Longitude and Latitude/Azimuth Nucleomagnetics Angle 33

    Note that Receivers Usually Do Not Extend Beyond Outer Repulsion Zone 35

    When Atoms Meet, the Repulsion Zones Operate Like Rubber Balls 37

Electrons Have Bonding in Open Paths past the Electron Field into the Nucleus at Specific Inclination Angles 39

    Two (2) Electrons Settle at the Nucleomagnetics Poles in Every Full Shell 40

    Electrons Settle in Balance in Opposing Hemispheres 40

    Understanding Bonding Angles for Carbon (109 degrees), Nitrogen (107), and Oxygen (104) 42

Electrons Have Positions at Specific Longitude/Azimuth Angles .. 44

    Using Electron 2c1 as the base X=0 Reference 46

Contributor and Receiver Assemblies Mimic Bonding Strength and Tension Using Springs and Engineered Magnets 48

Exterior Protons Occur with Contributing Hydrogens and Lone Protons 50

    Drives Exceptions to the Closed Space Ideal Gas Law 53

    Uniqueness of H2O Water (and other Contributing Hydrogen Molecules) 54

Molecules Depend on Electrons Holding Stable Settling Positions in Two Atoms 58

    Electrons Choose Which Atom Becomes Main 58

The Special Qualities of Water (H2O) 60

    AVSC and pH Levels – H_3O, H_4O, and So On 60

Building Layers of Chemicals 61

Challenges of Connector Distances Changing With Further Bonds 62

    Water (H2O) Proton Positions Changes from Alone Gas Phase to Multi-Molecule Liquid Phase 62

    When Carbon Change Bonding Angle Changes when Tail is Strong Halogen 63

Animating Chemical Reactions at the Particle Level 64

Conclusion 65

Endnotes 66

Electrons, for Most Engineering, Have Settling Positions

Engineering at the particle level is a challenge. Current teaching states that particles, like electrons, protons, and neutrons, do not have a knowable size, position, or speed (or the more you know one, the less you know the others); today particle physics is only statistics. That works great for some challenges, but for most people like me, we got into engineering because we like building blocks. Even if statistical, we need a best guess, a settling position, an engineering drawing, so we can puzzle real life challenges to workable solutions. We need to see to engineer.

Today, the position of a particles is a statistical 'cloud'. For one type of electron that distribution says that electrons might be in the white areas, but sometimes are in the orange and purple, so its statistical position estimate distribution looks like:

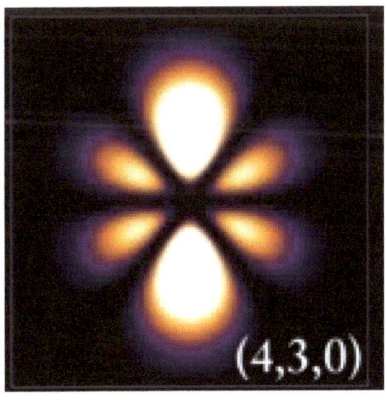

i

The great advance to science proposed here is that for many Chemical Engineering problems we can get back towards physical size, position, and speed. This model updates and replaces fully the current angular momentum model by adding, to the AVSC expanded-classical model, an electron-nucleon nucleomagnetics repulsion force to create a complete equilibrium theory. That provides the knowable forces, separate electrostatic force and the postulated nucleomagnetics (magnetic fields at the particle level) that determines how electrons settle into physical positions making the atom a stable geometric set. That determines, specifically, then known energy level (position) which create the AVSC Periodic Chart of Elements which describes the attributes, like electrical conductivity, bonding, and such of each Element.

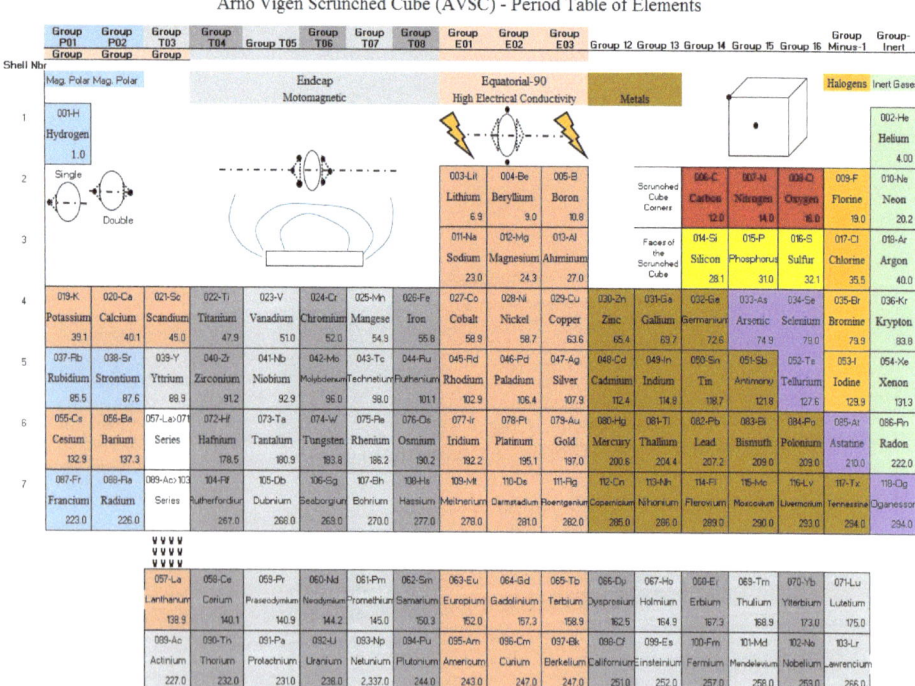

This provides the engineering reasoning for the attributes of the Elements:

- Outer electrons in the first column have one electron sticking out, along the nucleomagnetics axis. That makes this column very reactive.

- Outer electrons in the second column have two electrons (subshell-s in AVSC subshell-m for nucleomagnetics axis) in balance in opposite hemispheres, extending out along the nucleomagnetics axis.

- Up to three outer electrons for Elements in the middle of the Periodic Table, and some Elements on the left, settle at the equator, and that makes them easily exchange, hence these are the high *electrical conductivity* Elements. The outer electrons of 1, 2, or 3 are at the high-repulsion equator alone, and thereby highly likely to release into an electrical current.

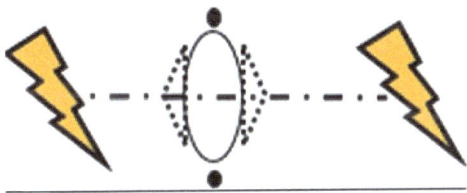

- Up to six outer electrons in the left fourth to eighth positions take a tight ('t') endcap position. That makes those oblong (oblate) which increases their *magnetic* properties.

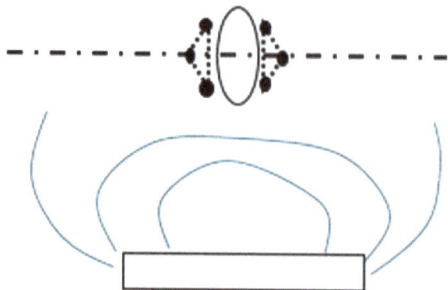

Most importantly, that combination makes electrons tend to settle into certain positions and distances which makes all the chemical properties knowable. Generally, the first two go to the poles of the magnetic field of the nucleus. Hence, we have Hydrogen and Helium, then Shell-1 is full. Every bigger electron shell also gets two electrons that settle into those preferred magnetic pole positions, so you have a more tightly bound subshell-s electrons (in AVSC it is called subshell-m for the nucleomagnetics axis settling position). Of course, there is statistics; yet for most problems there is an engineering method to discover solutions using distances, angles, fields, and knowable forces.

The Arno Vigen Scrunched Cube (AVSC) Atomic Model adds a nucleomagnetics force for every particle, and that drives the first two electrons in every shell into closer positions at the poles. Hence, we get 01-H Hydrogen and 02-He Helium only in the first Shell, and then we get two electrons (subshell-s) in every shell at tight positions (higher release energy) – that is, scrunched.

As I said, a picture is much easier for engineer types. For an atom of 10-N Neon, two electrons (2s2 which gets called 2m2 in AVSC) take the Shell-1 magnetic axis positions in blue. The either remaining electrons create a cube, with the two electrons at the nucleomagnetics axis 'scrunched' because the magnetics field strength is different there.

Electron Settling Positions in Scrunched Cube for 10-Ne Neon

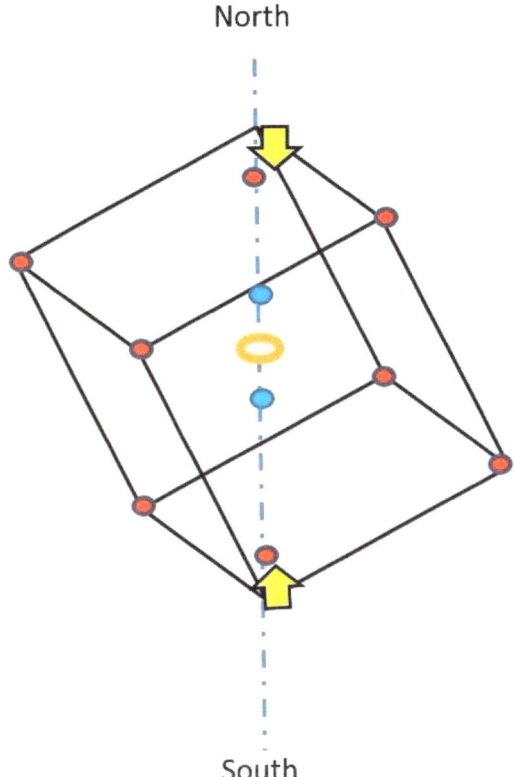

Please keep an open mind that the AVSC nucleomagnetics expanded classical model delivers a certain range of engineering solutions, even if your university had drilled into you all the discontinuities of time and space as the reality. Be patient. We will argue for a few decades about my updated theoretical model, but in the meantime, what us engineers need, and can use, is functional tools at atomic and subatomic distances:

- to engineer;

- to visualize atoms, molecules and bonds;

- to create animations of interactions at the particle level, particularly chemical reactions that can occur or not occur based upon particle level geometry, not just tables of relative electronegativity or statistics; and

- to calculate size, position, velocity, force, and their direction for subatomic particles within reasonable tolerances.

That said, I like build atoms, and molecules, and discovering in 3D how chemical reactions occur. These tools can do it better than statistical models – even if parts of the underlying theory get revises in the decades to come.

Big hugs. Let's get going.

3D Structures Build in Two Hemispheres of Settling Positions 1/3/5/7/7/5/3/1 from the Axis to the Equator to Axis

In AVSC, the subshells build form the axis to the equator to look, and back down again in the other hemisphere with electron counts in odds.

$$1/1$$
$$1/3/3/1$$
$$1/3/5/5/3/1$$
$$1/3/5/7/7/5/3/1$$

This odd-jumping adds to the squares for one hemisphere. The volume is layered so a volume of a circuit that is one electron deep.

$$1 = 1^2$$
$$1+3 = 4 = 2^2$$
$$1+3+5 = 9 = 3^2$$
$$1+3+5+7 = 16 = 4^2$$

1/3/5 Side View

One (1) in back

Two (2) in back

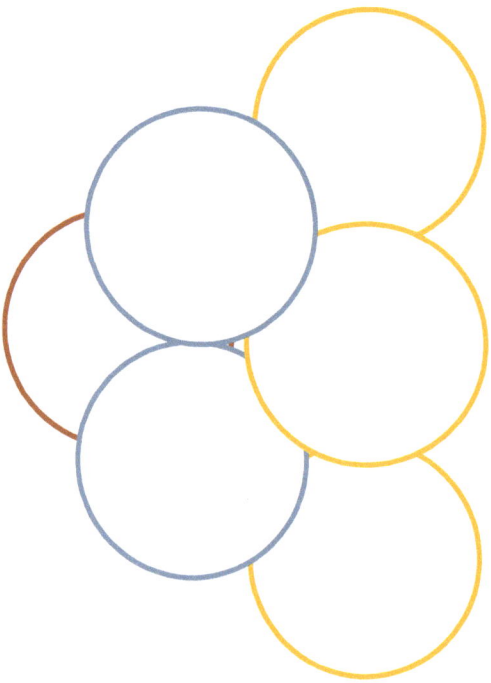

1/3/5 End / Polar View

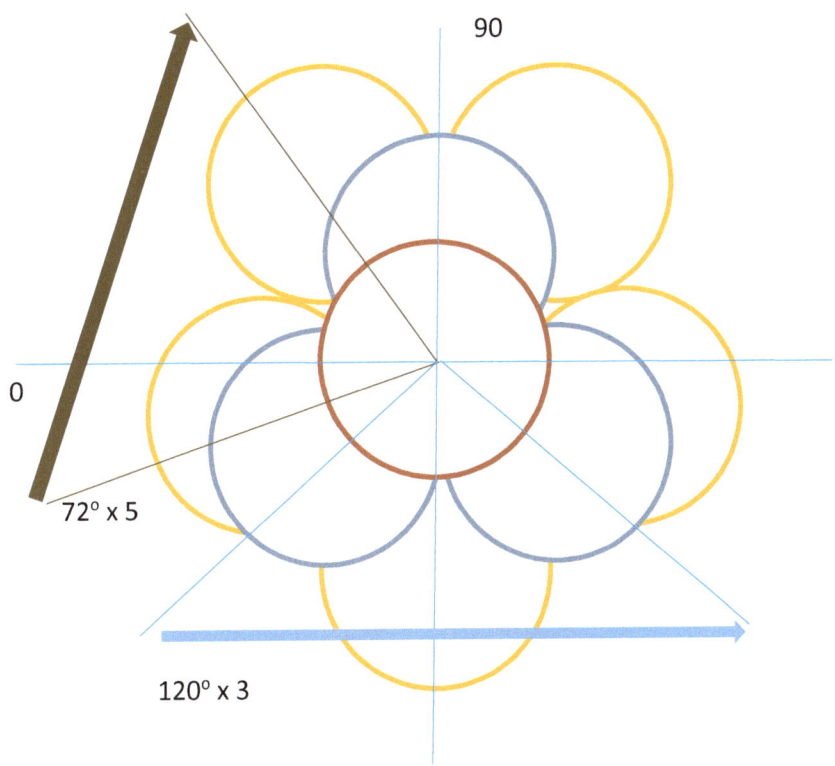

The latitude angles between the 3-per-hemisphere ring of electron positions is 360/3 = 120 (blue) degrees, and latitude angles between the 5-per-hemisphere electrons positions are 360/5 = 72 (blue) degrees.

When all positions not full of electrons, the empty position tend towards filling the axis first, building up towards the equator.

Finally, the empty positions in subshells like to get filled, at the lower bonding strength by outer electrons of other atoms. In a stable molecule, certain electrons settle into positions in both atoms to make the subshells full. Because of the repulsion of other electrons in that second atom, the position of that shared electron (not a pair) is usually further outward, but at the same angle to create a complete subshell.

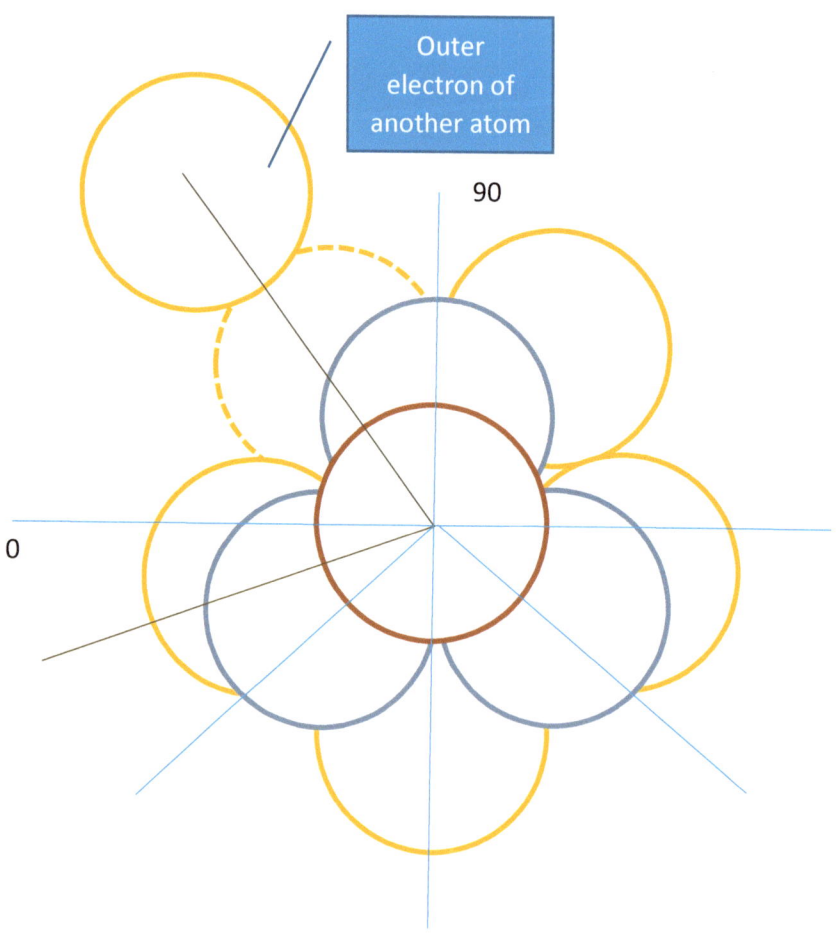

However, 2nd Layer Can Fit In Between One Layer to Create Repeats 1/3/3/5/5/7/7

If you look at more than one Shell, you get a different picture. There is room at the same inclination angle (a slice of the atom's physical structure perpendicular to the nucleomagnetics axis) for two layers at each count. A 2nd count of 3 can fit offset in between an inner layer of three. However, a 3rd set cannot because that would place the electrons back on top of (too close) to the electrons of the 1st layer of 3. It gets pushed out more such that that layer fits more electrons (5) as the most stable, efficient physical structure. The repulsive force from two layers in (yellow) keeps the structure from having a third layer of three (3) electrons.

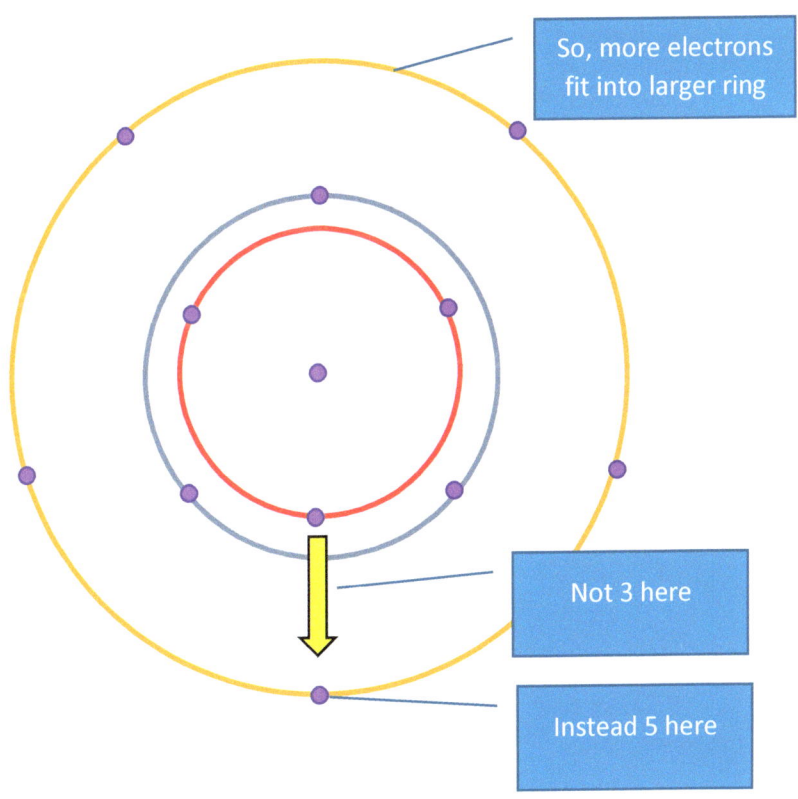

If you think of electrons, as done in AVSC CAD, as bubbles of repulsion the picture of the outer shells of a large atom look like:

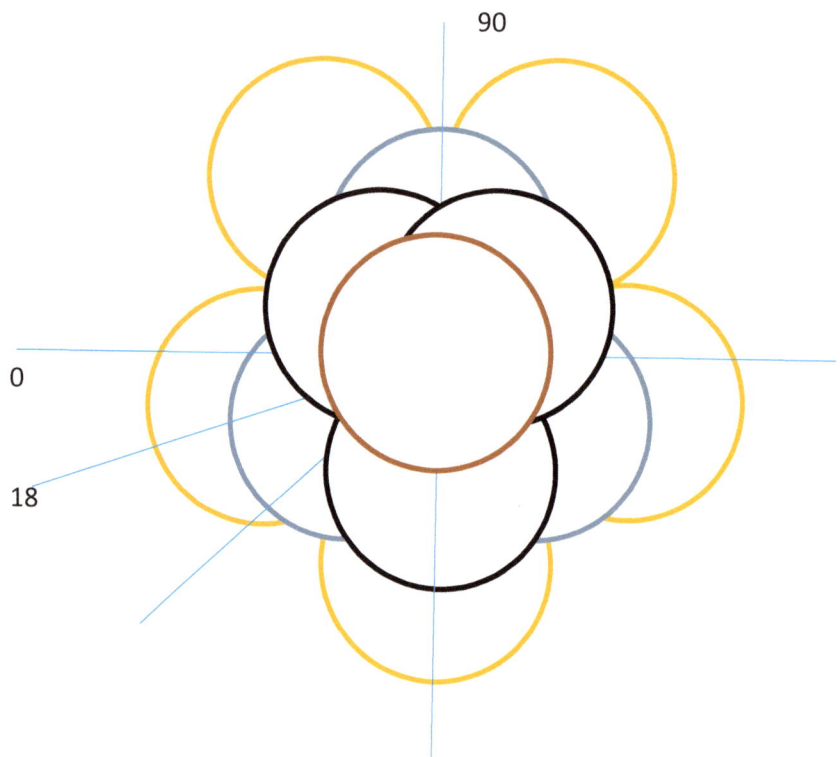

Multiple Shell Exterior Electron Repulsion Zone - End View

Shell-2 and Shell-3 Scrunched Cube

The basic idea of the Arno Vigen Scrunched Cube (AVSC) Atomic Model is that electrons fill at the balance of electron-proton electrostatic attraction and electron-nucleon nucleomagnetics repulsion. But, magnetics have that different shape, so as the nucleomagnetics field repulsion is weaker at the poles (that is different than a grand scale magnet), the first two electrons in every shell go to the poles. That makes every atom a Scrunched Cube for Shell-2 (corners of the cube) and Shell-3 (faces of the cube):

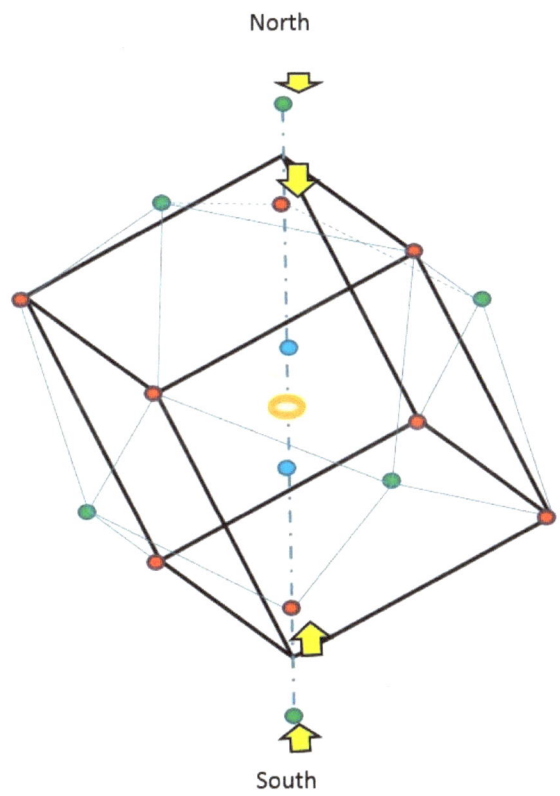

The first two electron go to the poles. Hence, Shell-1 is 01-H Hydrogen and 02-He Helium. Every other shall has two electrons at the poles and thereby closer. Knowing physics, that closer position is much more difficult to release (move to infinity).

The rest of the position build. The next stable structure in 3D is a 4-point tetrahedron – hence the 109.5 Carbon with one electron at the pole and three (3) electrons balanced

Of course, a tetrahedron does not have an axis through two points, so everything in AVSC gets build in two overlapping hemispheres (the physical model underlying the Pauli Exclusion Rule). And two overlapping tetrahedrons build into a cube – of course, again scrunched.

For Shell-3, the open positions are not the corners of the cube, but the six (6) size faces of the cube, with the two (2) electron that favor the nucleomagnetics axis, of course.

Equatorial Exceptions to Shell Filling

In AVSC, certain electrons will take 3D positions that are exception to the preference at the pole before equator general rule. First, electrons will take equator positions up to three in a new subshell. Hence, electrical conductivity/resistance in 1,2,3 ratio for Cobalt, Nickel, and Copper being the most well-known Elements for high electrical conductivity and low electrical resistance. Further, this explains that highest electrical conductivity for Copper (29-Cu), Aluminum (13-Al), Silver (47-Ag), and Gold (79-Au).

29-Cu Copper – Equatorial View

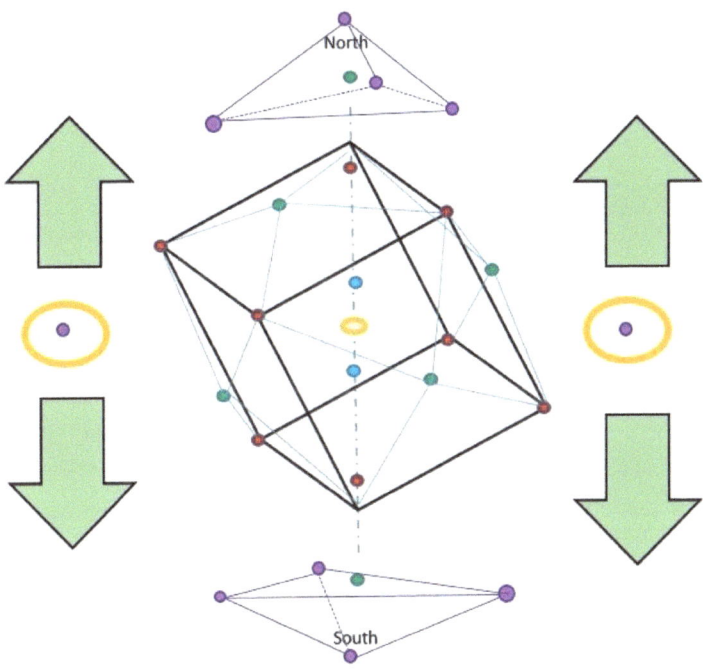

29-Cu Copper – Polar View

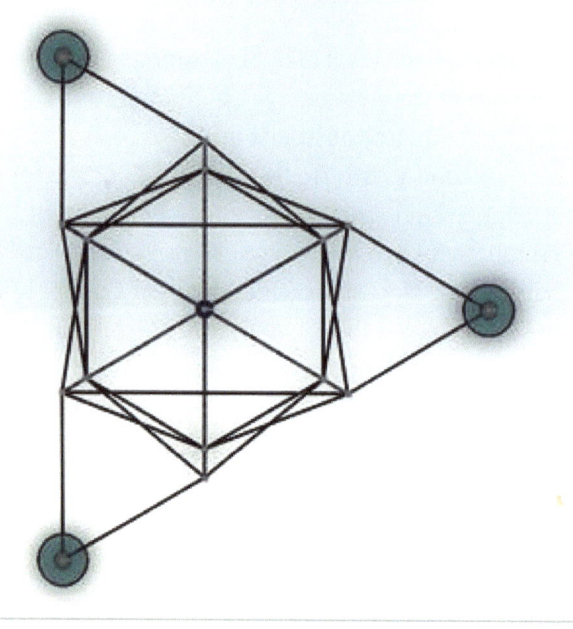

At the equator, we do not have Pauli exclusion, but can have three (3) electron in balance. This explains the electrical conductivity maximum in these positions.

Once a person understands this new method, every problem gets a deeper view which can engineer down to each particle. The rest of this book explains atoms, molecules, bonding, and chemical reactions with the specific of each particle position and

In AVSC Chemical Engineering Tools, Electrons (and Protons) have Repulsion Zones

Generally, an atom is a nucleus with surrounding electrons. Electrostatic force drives repulsion between the outer electrons; that is what makes atoms tend to remain the same element, and not change all the time.

Attached to every AVSC electron is a Repulsion Zone that provides an engineered method to make particles, of the same electrostatic charge, like electrons near electrons, to stay separated:

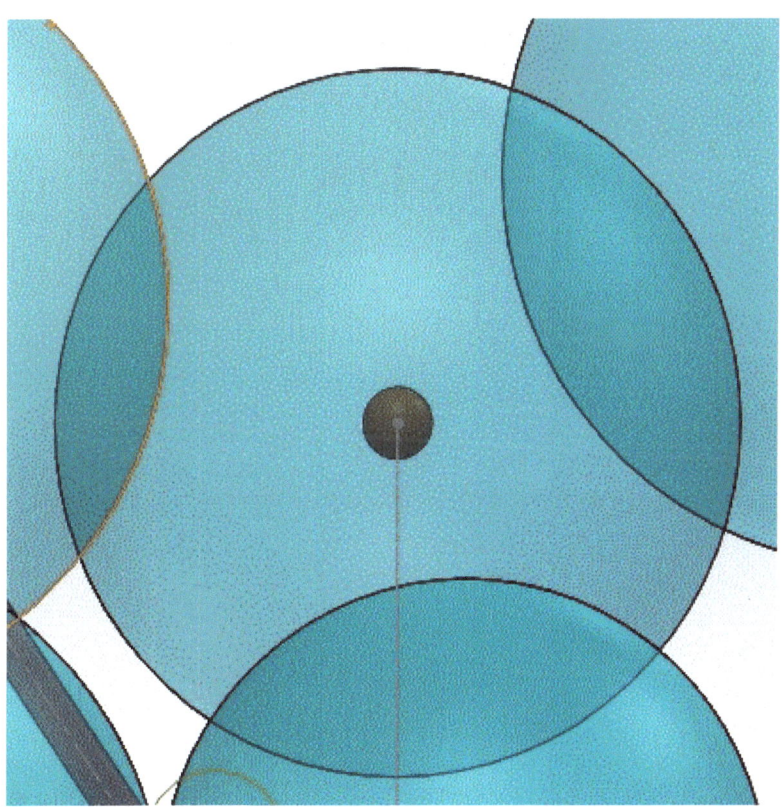

Repulsion Zones are engineered such that when the repulsion zones of two electrons meet they do not penetrate, but instead bounce off each other. In our preferred use, the material is translucent plastic which deforms slightly, and gives that acceleration back to the touching set of particles/fields.

When they meet in AVSC Animations, they bounce off each other.

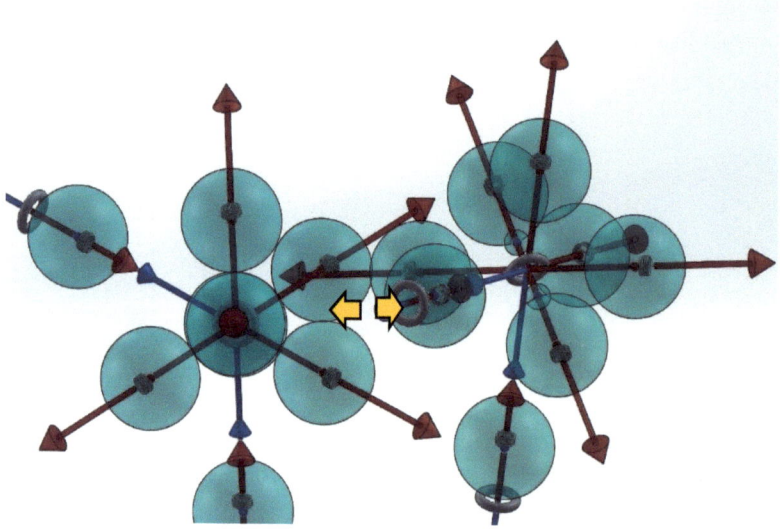

This mimics the way in which molecules, with exterior electrons tend to interaction - - - by bouncing off each other, keeping the atoms intact as a set.

Special Feature – Adjusting Repulsion Zones as a Set

Further, the size of all Repulsion Zones can get changed by a master directory .txt file. In that way, a user does not need to change particle by particle.

In the settings of an atom or a molecule, we place a file called 'Equations' which includes the statement:

"Evaluation Multiplier"= 1.5

This provides the way to change all the repulsion zones with one entry, and to keep the sizes in synch relative to each other.

In this method, you can go from evaluating an exothermic reaction or an endothermic reaction by reviewing if the repulsions zones block interaction at any one of the three levels (exothermic, natural-no-energy, and endothermic).

In endothermic reactions, a bond can occur by making the Repulsion Zones deform, like made of plastic, enough that the other molecules need to compress that repulsion zone in order to bond.

In exothermic reactions, a bond can occur with extra room between the repulsion zones, and the atoms will move together until the repulsion zones touch, creating the stronger bond, and releasing that extra energy.

In AVSC, by making the connectors overlap until the electron repulsion zones bump, that excess energy of the overlap gets released.

This distance measure in AVSC also shows why it becomes obvious why a bond with a low count atom, like Carbon, can bond tighter, and with extra energy, the repulsion zones are on the other side of the target atom, far away. The center of the repulsion zone of the approaching atom can move closer to the nucleus before touching (repelling) from the repulsion zones of the nearby electrons of 06-C Carbon. The yellow arrow is the overlap (exothermic energy).

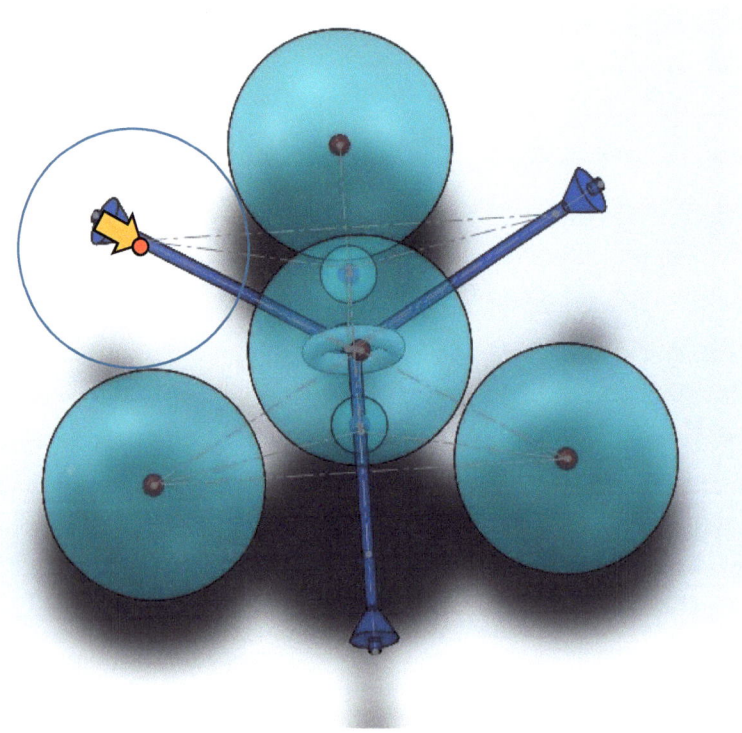

But, in larger atoms the outer electrons of nearby atoms create blockage to bonds, as seen in a 36-Kr Krypton atom that has not available open paths to create a shared-electron (covalent) bond.

In this case, the repulsion zone has lots of electrons nearby, so the chance that another atom's contributor can get to the receiver is nearly impossible. The yellow hashed arrow is the amount the other atom electron is pushed away from the basic bonding position, so the atoms cannot bond. The other atom repulsion blocks the electron from bonding with the required distance (radius = strength) along the few open paths to the nucleus form the exterior.

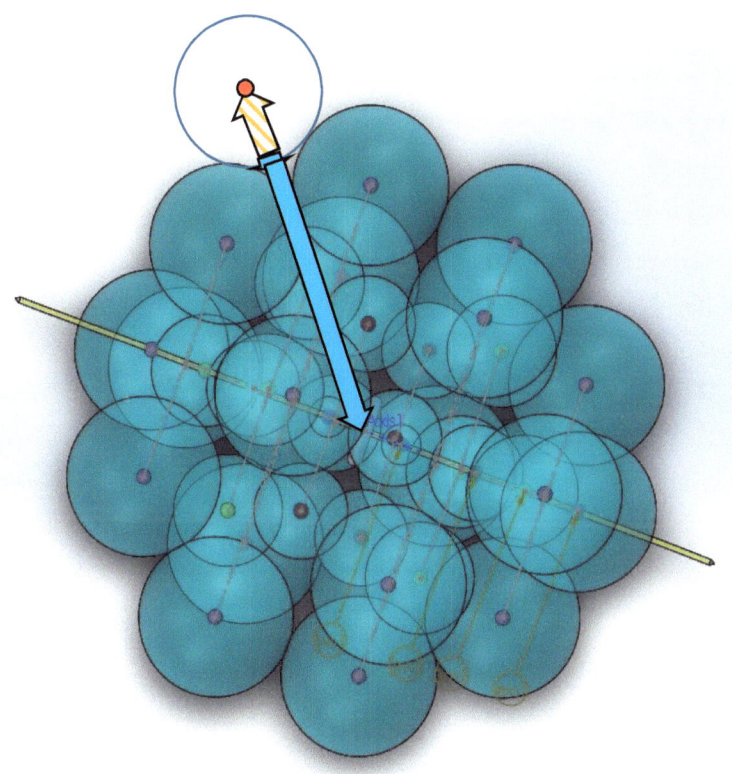

or b) the need to energy to get the bonding connectors without the exterior repulsion zones from hitting. Or more importantly, when an approaching atom has a longer connector, it can replace an existing bond.

The below example is that the Carbon-Oxygen, stressed by the spring angle, bond gets replaced in the presences of H+ protons onlys.

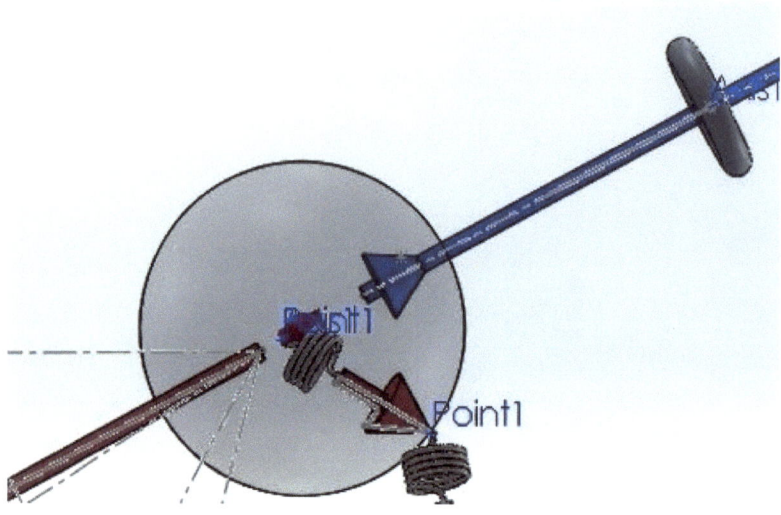

But does not react with a full Hydrogen because the repulsion zones bump. In the case of a full Hydrogen atom trying to break one of the CO2 double bonds, the repulsion zones form the other side actually bumps the Carbon electrons repulsion zones, and thereby the CO2 bond breaking reaction does not occur in just Hydrogen, only with H+ protons alone form the first picture.

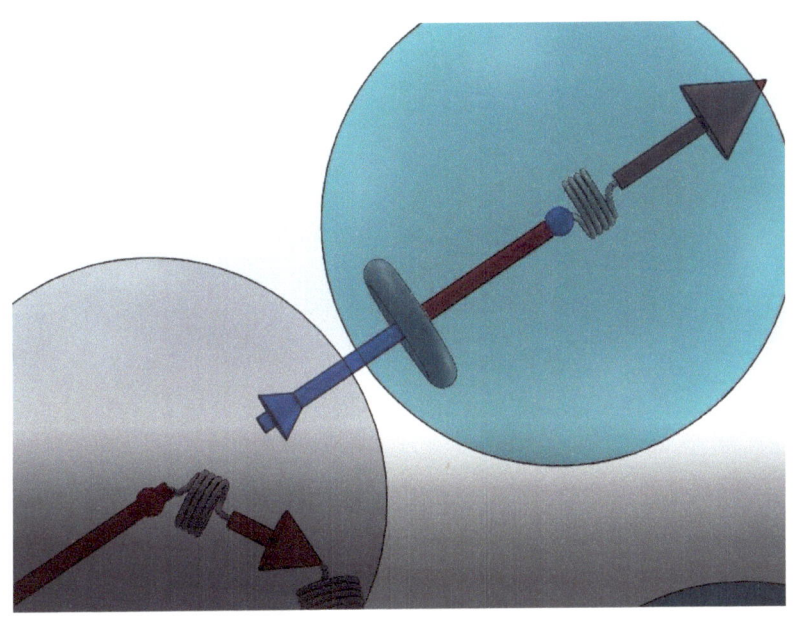

In AVSC Tools, Repulsion Zones Size Based upon the Distance (Layers) from the Exterior

Electrons repulsion strength further changes based upon the distance from the interaction, and since almost all interactions are at the surface, the strength of force, and thereby the size of the repulsion zones of interior electrons do not express as large.

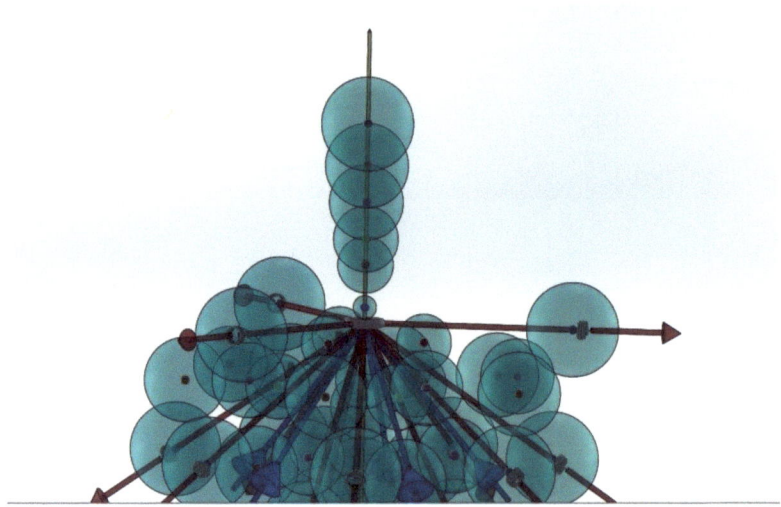

In reality, the electrons repulse each other equally, but the zone changes reflects that the impact on a potential bond is weighted towards the closer electrons because of distance, so those outer (closer to the other atom that might bond) have larger zones.

By using the translucent feature of the repulsion zones, a direction with multiple electrons darker, and even opaque.

For example, if you look at any large atom from the nucleomagnetics axis, there is no open path to the nucleus. Each shell has a subshell-m electrons so for Shell-6 there are six electrons in the path to the nucleus.

As one can see, there is rarely open path, receiving, or bonding along the nucleomagnetics axis. (Of course, these can be contributors – as in Magnesium or Sodium.)

Protons Have Repulsion Zones

As noted, in some reactions, there are external protons, and to distinguish a proton repulsion zone, AVSC will use gold as the tint for display.

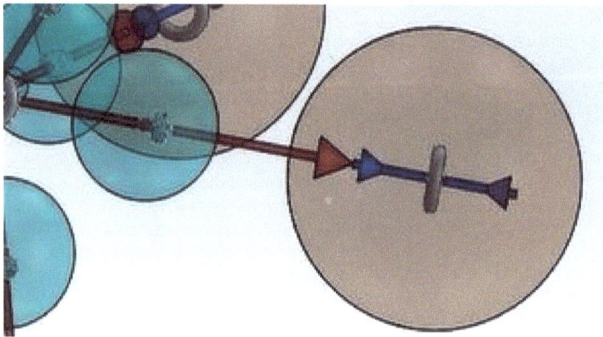

The most common use of this is water (H2O) which has two exterior protons versus six exterior electrons. Most atoms or molecules have only exterior electrons. That is the normal picture.

An Atom (Each Nucleus) Has Bond in Open Paths past the Electron Field into the Nucleus as Specific Inclination/Longitude and Latitude/Azimuth Nucleomagnetics Angle

The main engineering drawing of a bond is that a certain angles, the repulsions zones have open channels to the nucleus. Along those open channels, an electron from another atom can take that position, and create a new bonded molecule.

In AVSC, this presents as a receiver has an inward pointing cone to indicate it is receiving versus contributing.

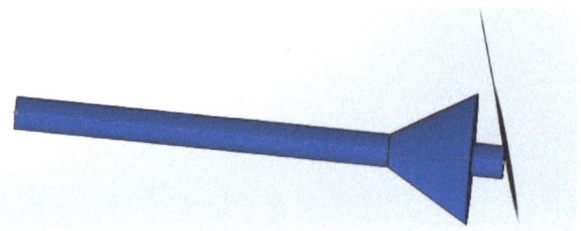

In AVSC Engineering and AVSC Animation, that open path gets filled, at the static nucleomagnetics inclination angle with a Receiver Assembly.

One end connects to the nucleus, the source of electrostatic attraction, and the other end is available to connect to a contributor of an exterior electron of another atom.

Note that Receivers Usually Do Not Extend Beyond Outer Repulsion Zone.

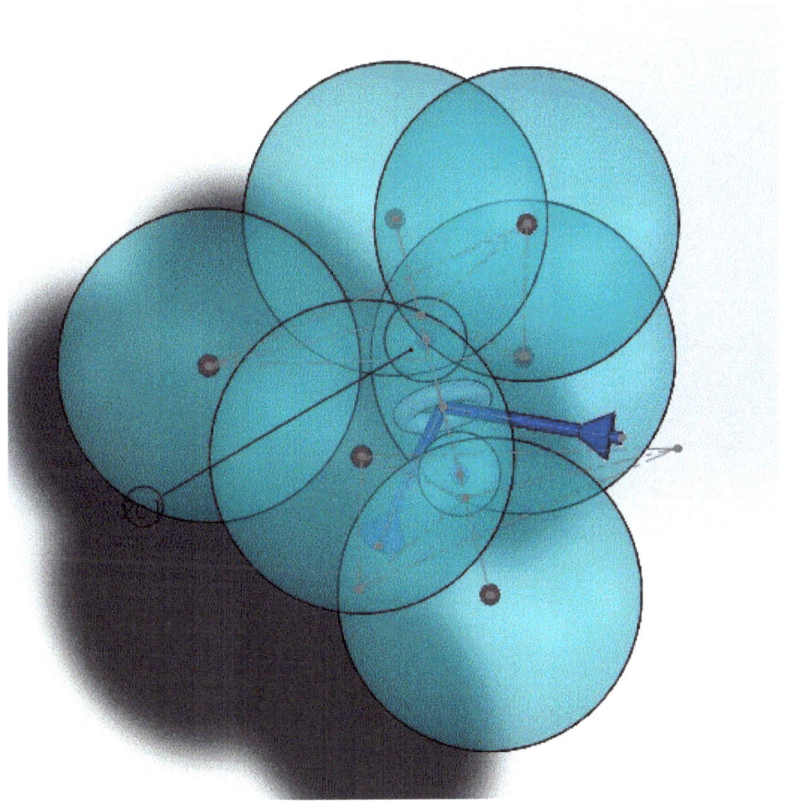

A bond needs electrons with extra spacing that give them a connector long enough to reach into the receiving atom without bumping into other repulsion zones.

In AVSC Engineering, we have separated the distance of covalent bonding between three basic segments. Atom #1 has a Receiver, Atom #2, has a distance from its nucleus to the outer electron which will form the bond, and between those two is a Contributor attached to the outer #2 electron.

Those distances of the three components are not equal. In fact, the most interesting aspect in 3D is that bonding electrons with their contributor must find a recessed receiver. If the outer repulsion zones are too close, or in blocking alignment with other electrons, then the bond is unlikely.

This is why many molecules that one might think possible or likely do not occur. In AVSC, we can engineer that failure in 3D graphics. You can see that atoms cannot fit together even if there is bonding position. (I will describe the 3D nature of AT bonds and GC bonds in DNA strand in a later example.)

In a simple example, a connector from Atom #2 must find a receiver in Atom #1 without Repulsion Zone blockage. This is done with the three segments connecting two atoms.

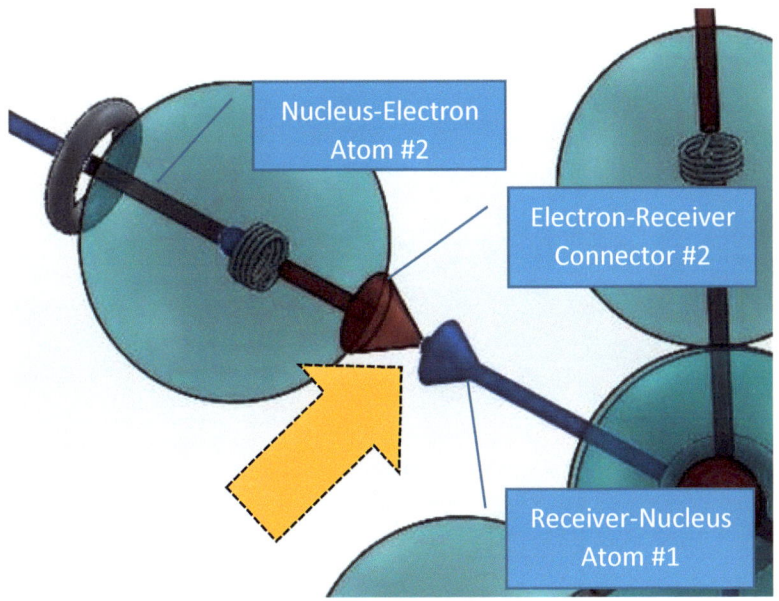

When Atoms Meet, the Repulsion Zones Operate Like Rubber Balls

On the other hand, sometimes the receiver does not find a contributor, and in that way, when repulsion zones meet, the atoms repel each other. In AVSC, repulsion zones are translucent rubbery balls around every electron particle.

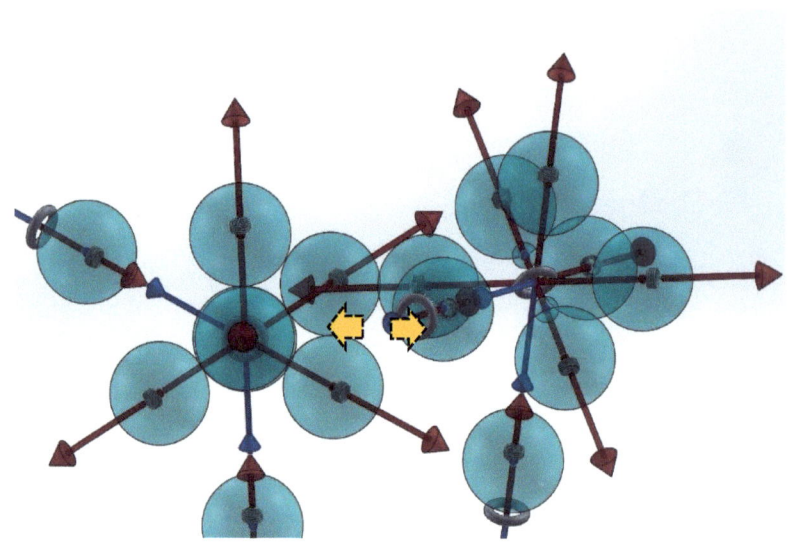

Electrons Have Bonding in Open Paths past the Electron Field into the Nucleus at Specific Inclination Angles

The structure is based upon a balance of forces, oriented, and driven by the strongest force at that distance range, which is the nucleomagnetics force from the protons and neutrons from the nucleus. In Oxygen, there are eight (8) protons, but, in the most common isotope, there are sixteen (16) nucleons (protons + neutrons). At the Bohr radius, electrostatic and nucleomagnetics are equal, but there are more nucleons, so generally more nucleomagnetics force.

Electron-electron bonds do not exist, so the electrostatic repulsion force of other electrons is never as much as the nucleus forces.

However, as a full set of electron in a lower shell, they have determine settling positions. For electron in the outer positions of larger shells, a) the nucleomagnetics force gets weaker faster ($1/d^3$), and b) the certain electrons are closer. a set of electrons in the subshell underneath creates a set of energy 'hills' that define a settling position more than the nucleomagnetics inclination angle.

Two (2) Electrons Settle at the Nucleomagnetics Poles in Every Full Shell

The shape of nucleomagnetics is slightly different. It is always repulsive, at both poles, and that repulsion is consistent based upon the nucleomagnetics inclination/longitude angle. Further, the repulsion is less at the poles, and more at the equator. That makes the two electrons in each shell settle on the nucleomagnetics axis.

Electrons Settle in Balance in Opposing Hemispheres

Everyone has heard of the Pauli Exclusion Rule. It says that electrons are in pairs, but their oriented 'spin' is opposite. There can be two, but there have opposite 'spin'. In AVSC, that is understood by the 3D nature of one electron on one side, in one hemisphere, and the Pauli-paired electron in the opposite hemisphere at the corresponding inclination/longitude and latitude/azimuth. In wave analysis, one electron in one hemisphere will move right, and the other move left, so the two opposite electrons act in opposition. One moves 'up' means that the other must move 'down' physically as well as 'spin' metaphorically.

However, note that in AVSC, the electrons are NOT shared in the actually bonding position. Instead, one atom has an electron which is pair, and the second has an open position with its Pauli-paired electron way over on the far side.

In AVSC, electrons in 2D are shown as in opposite positions, and bonding positions are shown as open circles. All this is done relative to the nucleomagnetics axis.

Equator View: Outer Subshell Structure
07-O Oxygen

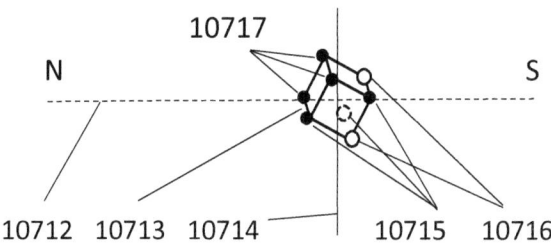

10712   10713   10713   10715   10716

Understanding Bonding Angles for Carbon (109 degrees), Nitrogen (107), and Oxygen (104)

the AVSC model address all bonding angle, including the Carbon, Nitrogen, Oxygen, by calculation because Nitrogen has an extra electron on the nucleomagnetics axis which pushes the tetrahedron into a different position.

Two tetrahedron that overlap are a cube. Yet, in AVSC, those are scrunched at two points, so the angles move from 109.5 for pure tetrahedron, Carbon, to smaller 107.5 for Nitrogen, to 104.5 for Oxygen, and so the different bonding angles are explained by the 3D geometry as determined in AVSC.

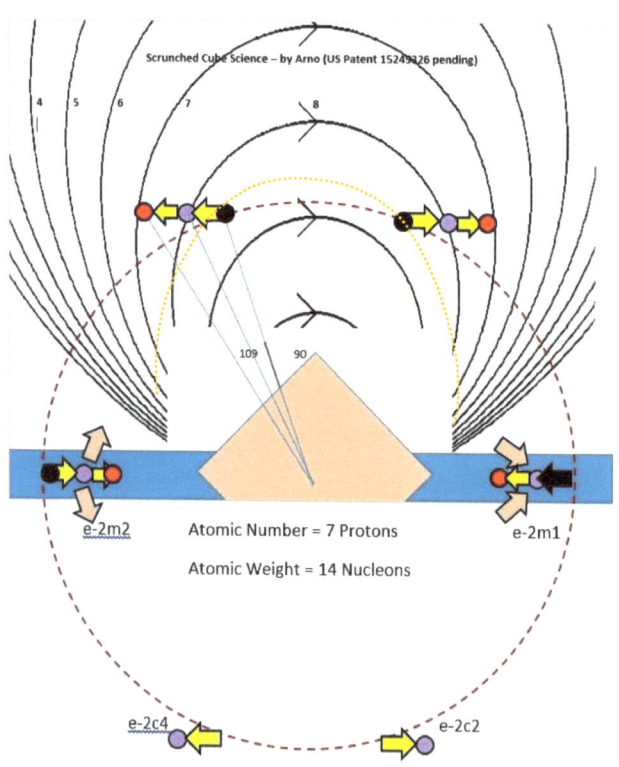

Remember that the number of nucleons grew from Carbon to Nitrogen from 10 to 12, so that magnetic field shape also grew accordingly. That pushes the settling position outward, but the outwards force is more towards the poles versus toward the equator.

The yellow hashed line of nucleomagnetics repulsion strength for 06-C Carbon (black) atom becomes the #7 line in the base drawing for 08-O Oxygen electron (red). The electron in Oxygen prefers moving outward from the equator, not upward. It has a balance of more attraction, but more repulsion at a different nucleomagnetics angle 66 decrees versus 70-1/2 degrees. That new balance makes the two bonding positions actually closer together at 104-1/2 degrees versus 109-1/2 degrees.

Electrons Have Positions at Specific Longitude/Azimuth Angles

We started with a picture of Outer Shells:

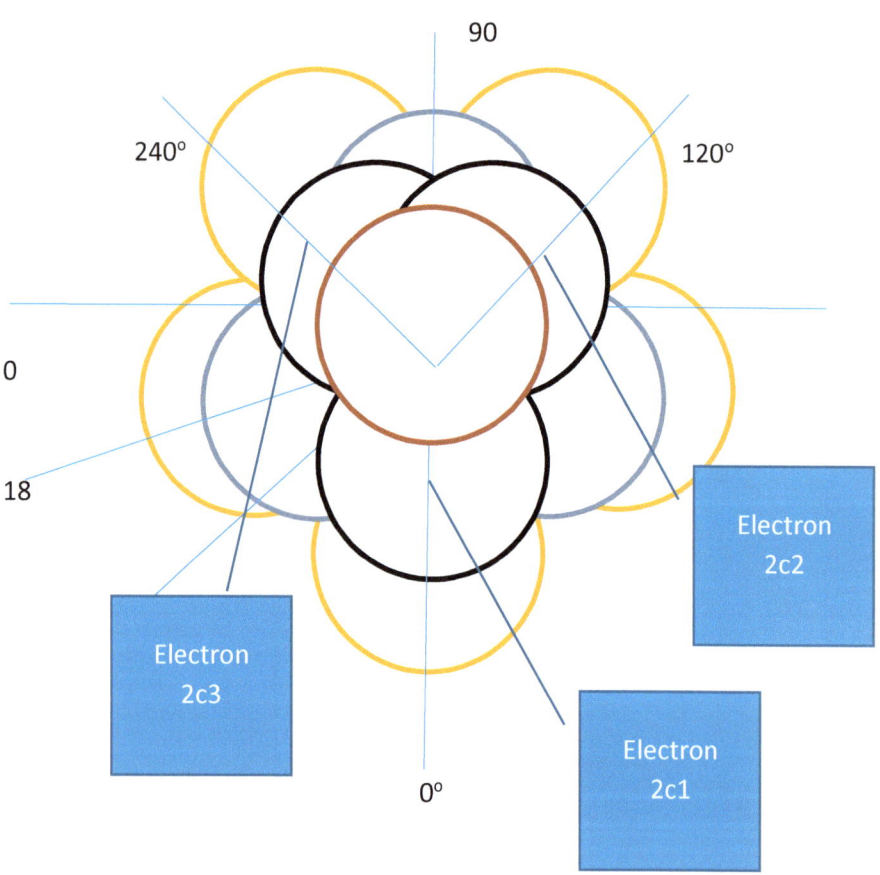

In the opposite hemisphere the position flip top and bottom, so the electrons in a 3-electron are 60, 180, 300.

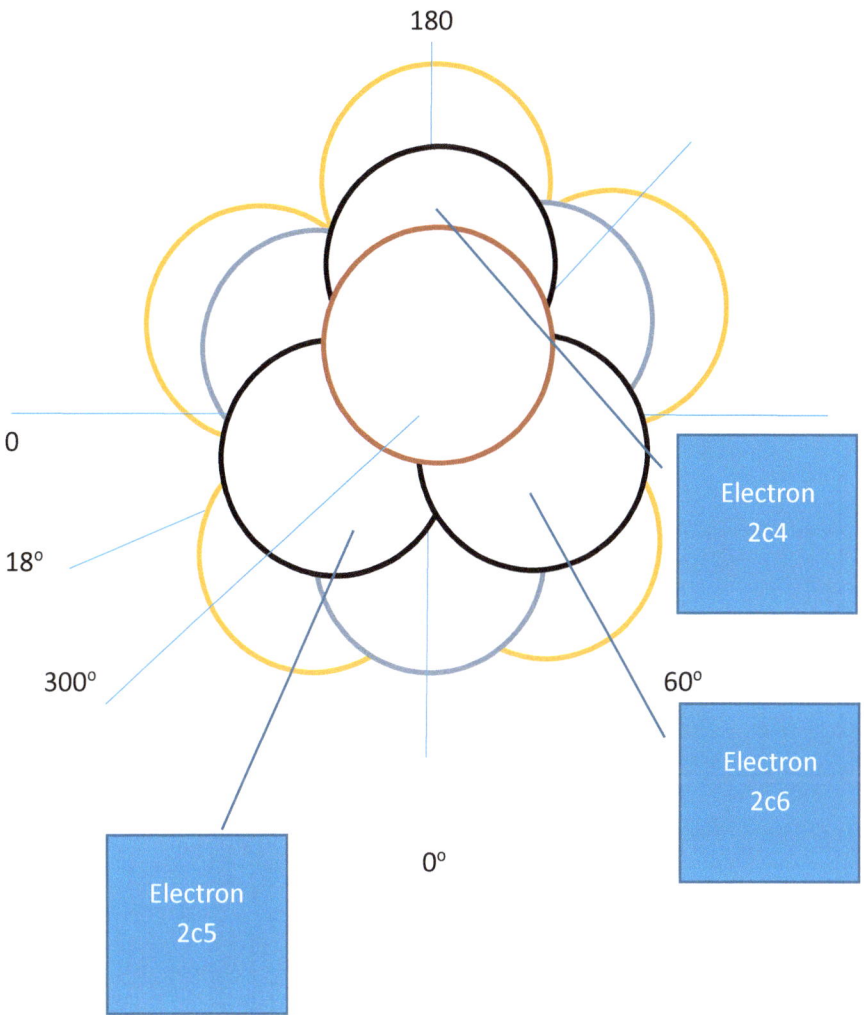

Multiple Shell Exterior Electron Repulsion Zone - End View

The choice of starting points could be in either hemisphere, but for consistency, we start number in the most filled end, and bonding occurs first in the opposite end. That is an Oxygen bond is set at electron 2c5 and 2c6.

Using Electron 2c1 as the base X=0 Reference

In AVSC Tools, all references start from the Layer-2 Electron position 2c1, which gets chosen as at x=0, so on the y-axis, and at Longitude Angle = 0. That is for engineering reference point. All positions in the same subshell have the same angle for inclination/longitude, and the same energy, so the same radius from electron to the nucleus.

Further, electron fill the '1' hemisphere first. That means bonding positions generally are in the higher numbers. So, an subshell filling table would look like below, filling the two axis positions, then the rest:

08-O Oxygen

Hemisphere	Inclination / Longitude Angle	Latitude Angle	AVSC Naming	Prior Naming
First	0	n/a	e-2m1	e-2s1
First	109.5	0	e-2c1	e-2p1
First	109.5	120	e-2c2	e-2p2
First	109.5	240	e-2c3	e-2p3
Second	0	n/a	e-2m2	e-2s2
Second	109.5	60	e-2c4	e-2p4
Second	109.5	180	b-2c5	b-2p5
Second	109.5	300	b-2c6	b-2p6

Please note that AVSC use the general formatting similar to existing chemistry training. Positions are based upon:

\#   Shell (number)

z   Subshell (letter)

\#   Electron position within the subshell

That means that 1m1 comes from 1=1st Shell, m=magnetic axis, and 1 means the first reference electron. 2c4 comes from 2=2nd Shell, c=corners of the scrunched cube, and 4 means the fourth (4th) reference electron; further since the off-axis of the cube (8-2) has six electrons, 1, 2, and 3 are in one hemisphere, and 4, 5, and 6 are in the other hemisphere.

Finally, the positions has a designator in the front to indicate either an electron ('e') fills the positions or it is a receiving positions for the bonding ('b-'electron from another atom.

e-   Electron Position

b-   Bonding Position

In AVSC, the full subshells are

Contributor and Receiver Assemblies Mimic Bonding Strength and Tension Using Springs and Engineered Magnets

There are more ways to mimic the engineering challenges of bonding. In Particle Level Chemical Engineering CAD, electrons are not locked into each other with these bars of connectors. Instead, there are various strength at which they can a) release and b) rotate. In AVSC, we mimic these by:

- Directional Connectors – using magnetic features to mimic the strength of the bond, and movements that break that magnetic connection. The connector is a red 'magnet' element S-N, and the receiver is a blue 'magnet' N-S such that the outer North connects only with receivers (N-S attracts), and not connectors (N-N repels).

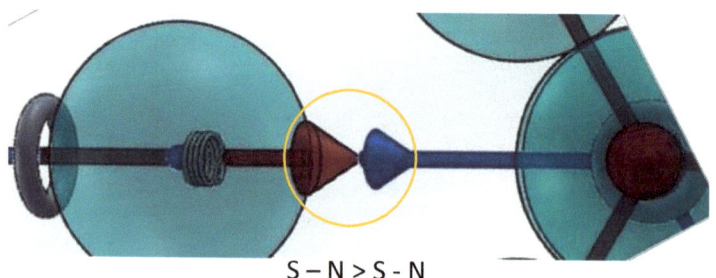

S – N > S - N

- Springs – using springs to allow bonding, especially double bonding along direction that do not align nucleus>electron> nucleus. However, after a certain tension increase, the bond swill break.

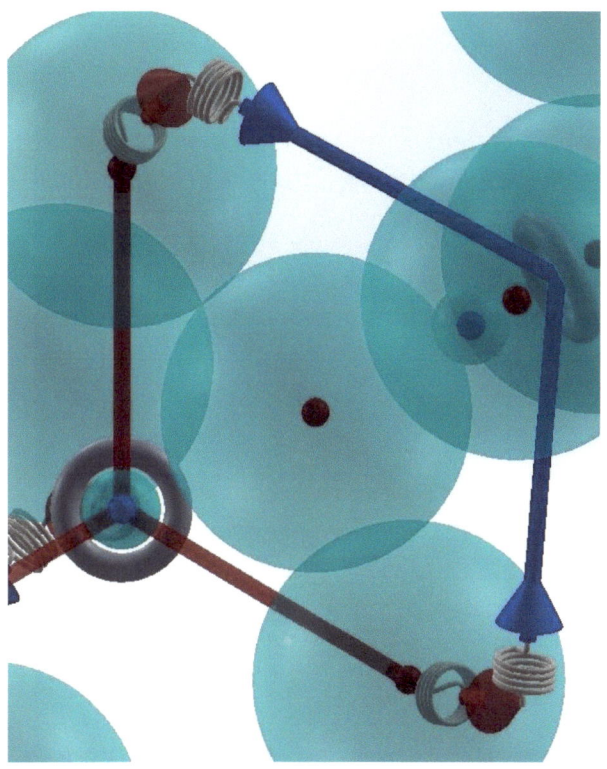

Note this is a 2D view of Carbon Dioxide (CO_2), and the electron(s) in the middle that looks like it is between the two is actually up and down; those are really out of the plane of the double bond by that same angle as double bonds angle are up and down. It is that 3D repulsion for those up and down electrons that creates the tension and settling positioning of the electrons in the double bond and connection such that two bond can go, but neither is a direct nucleus-electron-nucleus alignment of most single bonds.

The spring and connector in Engineering CAD can now mimic the actual forces that bond or break.

Exterior Protons Occur with Contributing Hydrogens and Lone Protons

Generally, an atom is a nucleus surrounded by electrons. That applies almost every situation; however, the exception is notable and important for chemical engineering.

A 01-H Hydrogen atom has only one proton in the nucleus and one electron. Further, that electron settles at the poles of the nucleomagnetics axis. As such, in the frame of reference of the atom, it is linear with an electron at one end, and a proton-nucleus at the other.

Most importantly, when a Hydrogen (01-H) atom is contributing, as in H2O water, the electron is inward, and the proton is outward. That creates the unusual situation where molecules where the exterior is not all electrons.

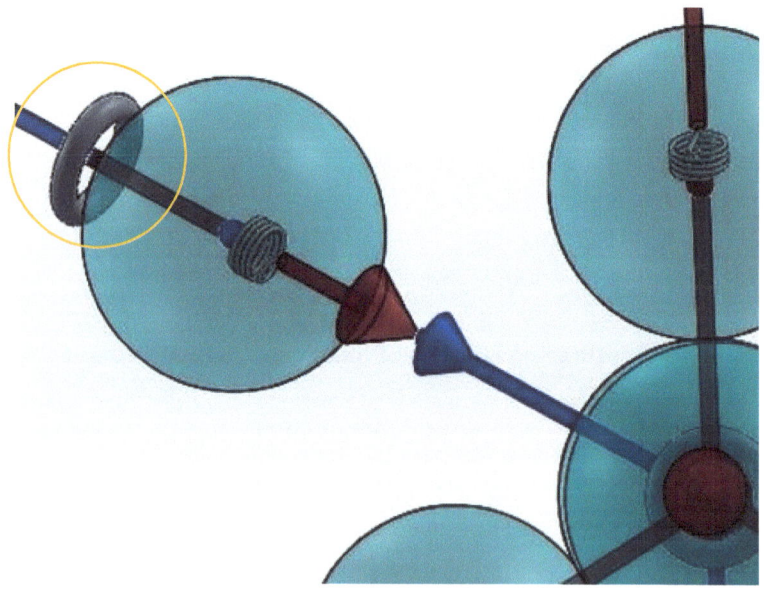

When a Hydrogen (01-H) attaches to its election of an atom as contributing towards the other atom in receiving structure, the proton is inward, and the electron is outward.

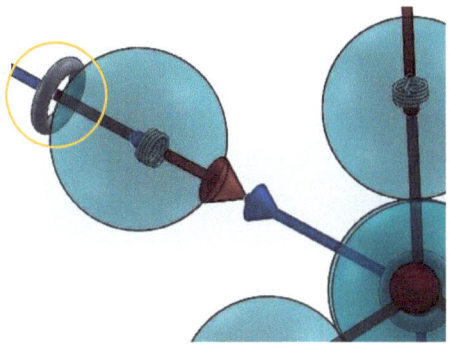

In AVSC Tools, we show the exterior protons with a different color (reddish) repulsion zone.

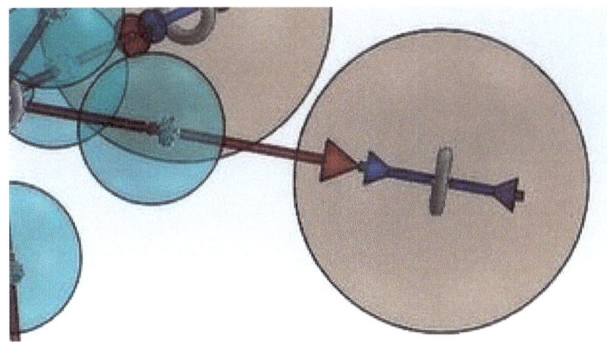

That creates an atom very different from the normal all electrons outward structure. Usually, electrons provide layers of repulsion zones such that bonding is difficult.

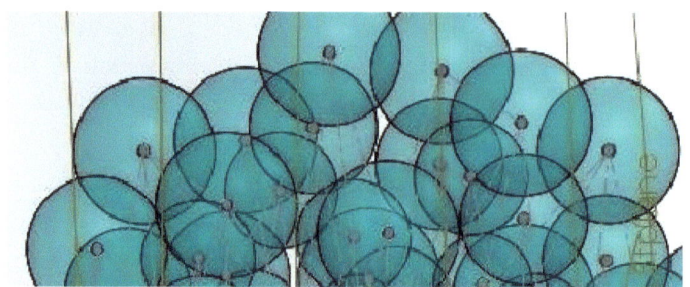

Drives Exceptions to the Closed Space Ideal Gas Law

Almost all atoms follow the Close Space Ideal Gas Law:

$$PV = nRT$$

P = Pressure

V = Volume

n = number of molecules

R = Constant

T = Temperature

That makes pressure go up linearly with Temperature, and temperature decrease as the volume is increase.

In that, once rotating, the extra negative electrostatic change in every direction becomes ubiquitous. In AVSC, every molecule is a structure with positive nucleus with electrons at the same average distance from the nucleus. That makes the calculation this linear relation, and all molecules give the same spacing in 3D.

In that way, the exteriors of atoms never form bonds, so the distance becomes dependent on the closed container volume, and you can calculate the pressure, volume, and temperature in ratio to the number of gaseous molecules.

The general rules are:

- Exterior of atoms are all electrons

- All atoms in gas state rotate fast enough that no bonds can occur (open paths close faster than any other molecule can approach to bond)

As a result, all gaseous electron-outward molecules generate the same surrounding space (at the same temperature, pressure, of course). A Neon is not different than a Di-Oxygen molecule.

However, that calculation is changed if the entire exterior is not electrons. If there is a proton-exterior, then the atom may never bond (be a gas), but the force of the electrons in the exterior will get reduced by the percentage of the sphere which is positive charged instead.

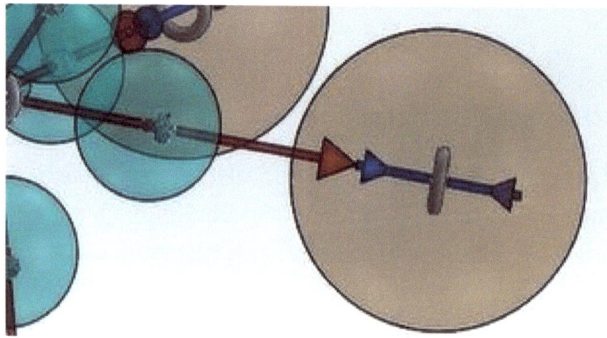

The visual AVSC picture helps to engineer the calculation of that

Uniqueness of H2O Water (and other Contributing Hydrogen Molecules)

Water (H2O) has six external electrons and two external protons (from the two bonded Hydrogens).

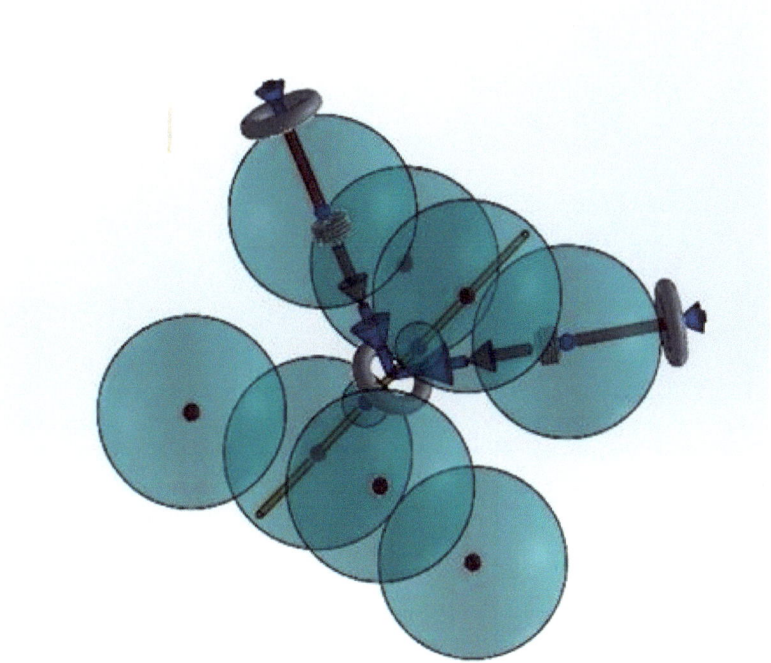

This CAD structure allows the Hydrogens (ring with blue receiving connector) to attach to Oxygen exterior electrons (red contributing connectors).

These are commonly called Hydrogen bonds. And the positions (distance) gets adjusted based upon the forces; other exterior electrons a) attract the Hydrogen proton, and b) repulse outer electrons, so that the bond length changes (which in turn changes the dipole electrostatic moment). This is the green expansion.

Other atoms also have this Hydrogen proton-outward positioning which leads to various positions for Hydrogen bonding wherever there is a Hydrogen in the contributing orientation.

For example, a Carbonic Acid molecules also has this proton-outward configuration on the Hydrogen-contributing to the Oxygen. That makes the molecule calculate as non-electron-outward atom exception to the ideal gas law. It makes the acid dissolve, especially in similar water.

Molecules Depend on Electrons Holding Stable Settling Positions in Two Atoms

The basic rule is that when one atom has open positions in a subshell, it generally accepts that an electron of a second atom takes that bonding position.

Electrons Choose Which Atom Becomes Main

One of the first chemical reactions in AVSC is the salt of a Halogen with an Alkali Metal, NaCl. The reactions of this type are called 'ionic' bonding. In liquid phase, the electron is more strongly attracted to the Chlorine (17-Cl) and thereby, it completes that subshell, transferring from the Sodium (12-Na) atom. The Sodium electrons are gone, so the Sodium now has the structure of a 10-Ne Neon full-shell, atom.

Ionic bonding is full transfer electrons, not sharing. In AVSC, the length of the receiving side (B + C) of the received is measurably less than the contributing distance (A).

Normal – Covalent Bond

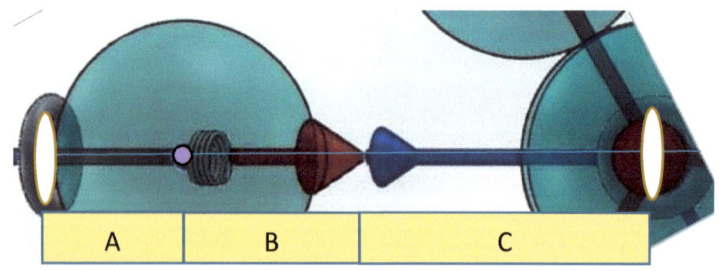

A < B + C

Ionic Bond

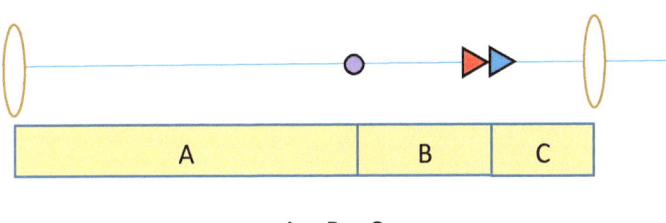

A > B + C

In this case, the electron tends to stay with the right atom when in liquids (dissolved). The right atom tends to become an ion (-), leaving the left atom as a cation (+). In visual engineering, a smaller distance means a stronger attraction, and the greater distance means a weak attraction. In this picture, the smaller distance is at the right means the electron stops its primary bond as A, and makes its primary bond the B/C connection. In AVSC CAD, the lengths of the connection substitutes for the relative calculations.

In the above example, the engineering CAD would actually chang from the bonding of two connectors at the right, to the connectors now being on the left, longer section

Ionic Bond – Electron Priority Changes to other Atom

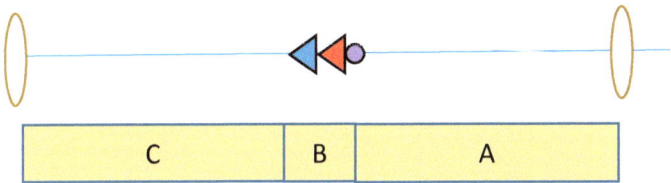

The Special Qualities of Water (H2O)

AVSC and pH Levels — H_3O, H_4O, and So On

Further, in low pH, the H2O which normally both contributes (6 Oxygen electrons) or receives (2 Hydrogen proton-outward structures), and change with extra protons (H+) fill up the Oxygen exterior electrons until the structure is all proton-outward. This makes reactions needing hydrogen-proton catalyst react much better and faster.

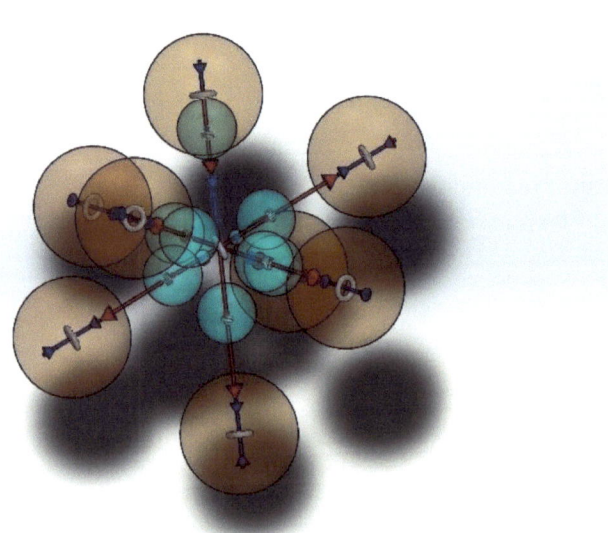

Building Layers of Chemicals

In AVSC CAD, we engineer forces to create assemblies for each atom or molecule. However, current CAD programs, like SolidWorks, overcalculate force until done forever.

In this way, we find that for strata engineering work, the conversion of the assemblies to parts, and removal of internal forces makes molecules build where you place them.

Otherwise, in CAD, you get layers that twist up at the edges based upon the CAD program wanting to extrapolate that forces multiple times such that the ends are twisted up.

Graphene – CAD Built

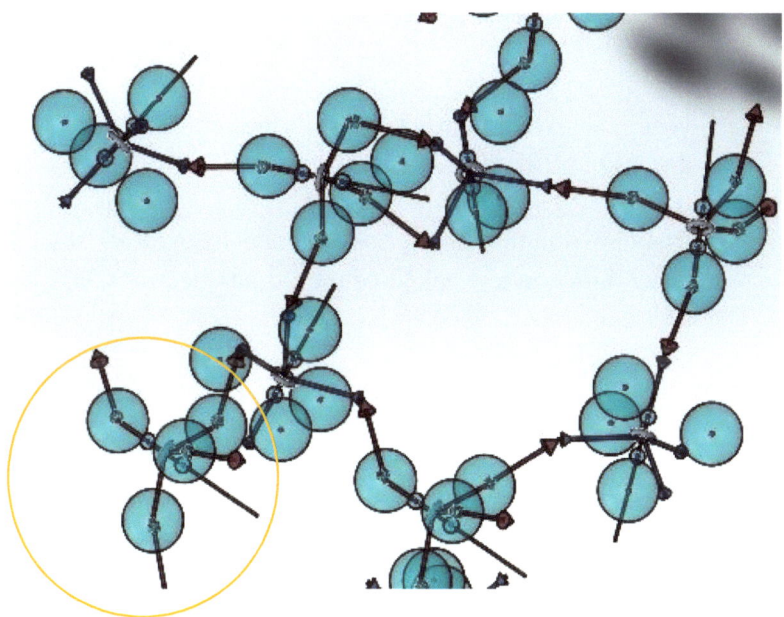

Challenges of Connector Distances Changing With Further Bonds

The main question in this project is when and why do we need to animate. The basic set of atoms provides the magnetics axis, the contributing and bonding positions, and their strength (as described by their distance).

However, the user must make some adjustments based upon geometry, atoms, and molecules so involved.

We already saw this with exothermic and endothermic reactions, but the structure was able to account for this by the distance needed or extra in the connectors once assembled.

However, the great challenge is when you have bonding in multiple directions to different types of atoms:

- When the exterior particle must further interact with groups of other molecules.

- One direction to a strong open-channel halogen, versus other direction to weak force from other electrons, like Hydrogen.

Water (H2O) Proton Positions Changes from Alone Gas Phase to Multi-Molecule Liquid Phase

One of the challenges of chemical engineering is that molecules act differently because of the atoms around them. Even so far as the molecules around one molecule.

Water as a gas has an electrostatic dipole moment, the center of the distribution of positive charge particles is different than the distribution of the negative charge particles. But, for water in gas phase, the dipole is 1.85 debye (D) which is a space of 6.17x 10^{-30}; yet, for water in liquid phase, that dipole is observed to change to 2.4 debye.[ii]

When Carbon Change Bonding Angle Changes when Tail is Strong Halogen

The idea that all bond lengths and angles are exactly the same is incomplete. Surrounding forces.

In the following carbon chain, each carbon has a different bonding angle based upon the closeness of electrons of atoms that bond. This is a chain of Carbons, but one end has a halogen, 36-Br Bromine, which has lots of electrons near the receiving bonding position. On the rest of the Carbon positions, the other atom is a Hydrogen, which has its electrons at 180 degrees, as far away as possible.

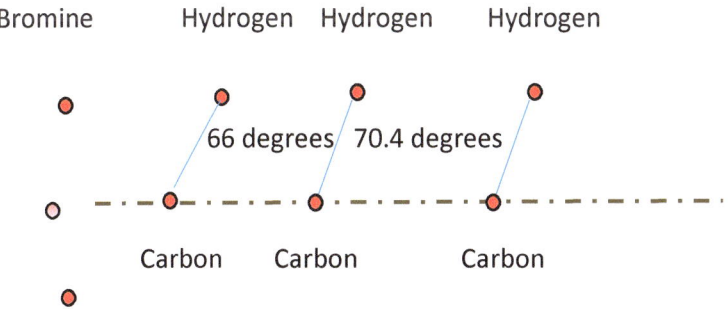

Animating Chemical Reactions at the Particle Level

The main question in this project is when and why do we need to animate. Most of what we do is engineer work without specific particles, and further, the majority of what happens is millions of atoms and molecules, so we cannot detail these one by one at this level of particle-level chemistry engineering.

- Many problems are solved by large quantity statistics.

- Many problems are solved by simply knowing the bonding angle of the relative molecular combination of atoms.

Yet, for certain problem, a statistical method or just bonding angle do not give the correct solution. In those cases, a direct engineering of what happens to each particle yield insight needed to explore ways to innovate.

Conclusion

There are millions of derivative engineering work with the easy tools of AVSC. With these tools, chemistry is now a workable structure that can use generally traditional engineering CAD exactly when the simple solutions cannot provide answers. This presentation is just a tiny part of endless potential. I am sure you have ideas and questions. Please join me in a great adventure to advance science understanding and application.

Big hugs, let's get started.

Arno

Endnotes

i
https://upload.wikimedia.org/wikipedia/commons/e/e7/Hydrogen_Density_Plots.png

ii https://msu.edu/course/css/850/snapshot.afs/teppen/physical_chemistry_of_water.htm

www.ingramcontent.com/pod-product-compliance
Lightning Source LLC
Chambersburg PA
CBHW040234220526
45473CB00001B/242